TECHNOLOGICAL EMANCIPATION OF NIGERIA
- The Role of Chemical Engineering
(A Collection of Lectures)

By

B. N. Nnolim, B.Sc.(Lon), ACGI, M.Sc.(Lon), D.I.C,
FNSChE, FNSE
Formerly of the Departments of Chemical Engineering
Institute of Management and Technology, Enugu,
Enugu State, Nigeria
and
Nnamdi Azikiwe University, Awka,
Anambra State, Nigeria

First Published and Printed 1994
by
CECTA (NIGERIA) LIMITED
159 Chime Avenue, New Haven.
P.O. Box 1306,
ENUGU
NIGERIA

©   1994 by B. N. Nnolim

ISBN 978-2396-14-1

Republished 2009

by

Ben Nnolim Books
7 Sandway Path,
St. Mary Cray,
Orpington, Kent.
BR5 3TS, UK

ISBN 978-1-906914-18-9

TABLE OF CONTENTS

	PREFACE	7
CHAPTER ONE	TECHNOLOGICAL EMANCIPATION OF NIGERIA	11
1.1	The Prospects for Technological Emancipation of Nigeria	13
1.2	Nigerian Engineering in the Context of Nigeria's Technological Underdevelopment	19
1.3	The Call for a New Generation of Engineers and Technologists	29
CHAPTER TWO	THE ROLE OF CHEMICAL ENGINEERING	39
2.1	The Chemical Engineer and Society	41
2.2	Nigerian Technology and the Use of Local Raw Materials - The Role of Chemical Engineering	45
2.3	The Place of Chemical Engineering in Raw Material Processing	56
2.4	Chemical Engineering Processes and Pollution	67
2.5	Chemical Engineering Prospects in a Depressed Economy	72

PREFACE

An intrinsic weakness of most tertiary level education is the lack of breadth and, sometimes, depth, in syllabus, necessitated by the need to address, quickly, the professional demands of a chosen discipline. The assumption is, always, that these issues, of perspective and panoramic coverage of subjects, would have been adequately treated at the primary and secondary education levels and that the graduate would widen his/her horizons by personal reading. Experience, however, has not been on the side of these assumptions and engineers, lawyers, doctors, accountants, economists, etc, graduating from tertiary institutions in Nigeria and overseas; continue to practice their professions as mere technicians and hired hands who lack the requisite worldview and motivation to make their profession count.

Recognition of this problem has resulted in the inclusion, in many countries, of a general study programme in most degree or non-degree tertiary level curricular. The idea is that scientists should know something about the humanities while the non-scientists should know a thing or two about science and engineering. Yet this has not solved the problem as lectures in the humanities are given to scientists and engineers by lecturers who have no training and very little understanding of science and engineering. Similarly, scientists and engineers, not versed in the humanities, are often given the task of introducing non-science and non-engineering students to science and engineering. The more developed countries eventually solve this problem by employing, in their organisations, the full plethora of professionals - doctors, engineers, lawyers, psychologists, historians etc. depending on their organisational goals and products. This practice is not, at present in vogue in Nigeria so that student training at the tertiary level has to be as complete as possible, before employment.

A very popular approach in Nigerian Institutions, therefore, is to make up the general study programme with Student Week activities in which among other things, eminent lecturers or professionals are invited to deliver lectures on selected issues of current topical interest to students in a given profession.

Thus, in a Chemical Engineering Week, Accountancy Week, Architecture Week etc, students are sensitised to the wider meaning, social standing and responsibility or merely to the successes or problems of their chosen profession in society. This has several advantages chief of which is the mental and intellectual preparation of the students for what they are likely to expect to meet after graduation.

After almost twenty five years of being a guest speaker on these occasions at the University of Nigeria. Nsukka, IMT Enugu, Federal University of Technology, Owerri, the Enugu State University of Technology and at other professional seminars, I have come to see, from the success of past students, that there is need for wider dissemination of these lectures, even if many of them are dated either by social, political or technological changes since they were first given.

In order to retain focus, I have excluded from this publication, lectures given specifically on career patterns and concentrated on more general issues likely to be encountered by engineering professionals. The hope is that these lectures will still find relevance among engineering students, especially, chemical engineering students in our universities and polytechnics.

I wish to seize this opportunity to thank the various Associations of Mechanical and Chemical Engineering Students at the IMT,

Enugu, Federal University of Technology, Owerri, Anambra State University of Technology, Enugu, Enugu State University of Technology, Enugu, and the Anambra State Ministry of Industry and Technology, the Enugu State Branch of the Nigerian Library Association, who invited me to deliver these lectures and, on some occasions, suggested the topics.

In almost all cases, the invitation to deliver these lectures always came at very short notice and only the most dedicated and conscientious typists and secretaries could have produced them in time. I have been very fortunate in having these typists and secretaries of which Mrs. Julie Eze, Mr. Fidelis Ugwu, Mr. C. Ozoagu, Mr. Matthew Agala, and Mrs. Catherine Nwadioke, all of the IMT, Enugu, stand out, for special mention and acknowledgement. Finally I thank God for giving me the energy to be able to prepare and deliver these lectures at very short notice.

B. N. Nnolim December, 1992.

CHAPTER ONE

TECHNOLOGICAL EMANCIPATION OF NIGERIA

1.1 The Prospects For Technological Emancipation Of Nigeria.

Being a Lecture Delivered at a Symposium organised by the Association of Mechanical Engineering Students of the IMT, Enugu.

Ladies and Gentlemen,

The invitation to deliver this lecture was received at very short notice. In spite of my very tight schedule, however, I accepted to give this talk for two reasons. One, technological issues, as they relate to Nigeria, have been my pet subject for many years. Secondly, I was very impressed by the maturity and focus which enabled the Association of Mechanical Engineering Students (AMES) to zero in so accurately on the key Nigerian problem - technological emancipation.

In examining the prospects of technological emancipation of Nigeria, I shall adopt a very simple approach. First I shall attempt to identify the ingredients of technological emancipation. Then I shall look for the presence or absence of these ingredients in Nigeria. Then I shall relate these to the prospects of technological emancipation.

The Meaning of Technology

But what is technology? Is it science or science based? Is it engineering or engineering based? Or is it one of the humanities? I am afraid that the answer is both none and all of these.

There are many definitions of technology in use today but I shall use only three of them in this talk. The first definition is a fundamental one and says that technology is the sum total of knowledge and experience, ie:

$$Technology = Knowledge + Experience$$

Fundamentalists will argue that knowledge is also a kind of experience since you cannot know something unless you have experienced it.

The second definition says that technology is a complex system of interactions between the history, culture, physical and resource environment, social, political and commercial objectives and practices of any society in a process in which it attempts or has attempted to solve its problems or sustain itself. As Patrick Moynihan, Democratic Senator of New York said in the 1970s *"Technology is not a product. It is a system"*.

The third definition says that technology is simply the business or process of providing goods and services to human beings in a convenient, affordable and acceptable manner.

Whichever definition you choose, I believe that you all will agree that technology is concerned with large scale solution of human problems. Any society, therefore, which can predictably and reliably solve its problems, whenever, wherever and however they occur, can then be said to be technologically emancipated.

Components Of The Technological Problem

It is my view that any technological problem can be decomposed into six problem units namely:- (a) the knowledge problem (b) the terminology or jargon problem (c) the tool utilisation problem (d) the design problem (e) the operation problem and (f) the management problem. A more business enterprise oriented analysis defines fourteen essential functions, divided into four groups, in the management of technology namely:- (Marvin, 1973):

1. The Creation Functions
   - Basic Scientific Investigations
   - Technological Forecasting
   - Business Conditions Projections
   - Demand Analysis.
2. The Co-Ordination Functions
   - Long Range Programme Formulation
   - Product Planning
   - Product Application studies
3. The Commercialisation Functions
   - Prototype Development
   - Pilot Plant Production
   - Process Development
   - Technical Services Development
4. The Capitalisation Functions
   - Patents and Licensing
   - Mergers and Acquisitions
   - Pioneer Sales and Marketing

Solution of these problems constitutes a technological solution and persistent expertise and success in these solutions is technological emancipation. Let us, now, consider the six components of the problem, according to my definition.

The Knowledge Problem

This involves the acquisition of basic information and understanding of the natural phenomena which govern the problem to be solved. Through the sciences, art, literature etc, this knowledge, as developed in all cultures, is now widely available to anyone who wants it. Fortunately in Nigeria, phenomenological knowledge is highly valued and is very popular in our primary, secondary and tertiary institutions.

The Terminology Problem

For knowledge to be useful, it has to be communicated. Certain terminology have to be developed for efficient communication in any knowledge area or discipline. The acquisition of relevant terminology usually precedes the acquisition of knowledge. This is not lacking in Nigeria.

The Tool Utilisation Problem

To apply the knowledge acquired in the solution of a given problem certain tools, not always hardware, must be developed or their use learned. Such tools may be mechanical or machines or methods of analysis. This is an area where there is much weakness in the Nigerian attempts at technological emancipation.

The Design Problem

This involves the translation of knowledge and ideas into three dimensional pieces of hardware within given or defined operating constraints. Extent of success here depends very strongly on the proficiency in tool utilisation and knowledge. Again Nigeria is very weak in this area.

The Operating Problem

This concerns the manipulation of the inputs and outputs of a given process within the constraints of a social or physical system or a piece of equipment. The inputs and outputs may be the investment in, and success of, an educational system or physical materials being processed in and out of a mechanical system. Again Nigeria has not scored highly enough in this area as can be seen in the performance of most public institutions in the country – National Electric Power Authority (NEPA), the Nigerian Police, national and state ministries of health, education etc.

The Management Problem

This involves the organisation of resources, finance, men, materials and machines, for the promotion and control of the other five processes towards the solution of the problem. Again, Nigeria has not scored enough to indicate progress in this area. The reasons are many - political instability, mediocrity in key places, corruption etc - products, in my view, of interactions of failure in the other component problem areas.

Conclusions

In summary, what are the prospects for technological emancipation of Nigeria? If the present score card is anything to go by, and continues for a number of years, the conclusion must be that the prospects for technological emancipation of Nigeria are very bleak indeed. However if attempts are made to change the score card, then there are bright prospects especially in view of the abundance of human and material resources of this country. What are these attempts that must be made?

- Nigerian professionals should attempt to acquire technical and analytical skills in addition to mere phenomenological knowledge which appears to be the main preoccupation of Nigerian educational, industrial and social systems in which Nigerians hold the posts and expatriates do the jobs or they are not done at all or well enough.

- Nigerian professionals should get more involved in design projects of consequence and not demonstration or news catching projects.
- Nigerian professionals should influence management decisions by being more authoritative and expert in their advice and recommendations to policy makers.
- Nigerian professionals should take stronger positions against government policies inimical to technological development in Nigeria.

Thank you.

References

Marvin P: (1973): Fundamentals of Effective R & D Management, ACS Audio Course, Am. Chem. Soc, Washington D.C., USA.

1.2 Nigerian Engineering in the Context of Nigeria's Technological Underdevelopment

Being a paper presented to the Student Chapter of the Nigerian Society of Chemical Engineers, Federal University of Technology, Owerri

Abstract

Although every human society has its own indigenous technology, it is often exposed, by choice or accident, to the best in world technology. Those societies, which take seriously the exploitation of this world technology, survive while those which do not, either remain technologically backward to bear the brunt of the affluence, imperialism and profligacy of the more successful societies or become extinct. African societies have, century after century, lost the opportunities for technological supremacy. It, especially Nigeria, has now another opportunity. It is our duty to ensure that this opportunity is not lost again.

Introduction

It is generally agreed that Nigeria is an underdeveloped, or to use a more popular term, a developing, country. This is in spite of its wealth of human and material resources. Fundamental services such as the reliable supply of clean water, electricity, food, social services, roads, etc. cannot be provided. Nigeria has not been able to develop or sustain an adequate means of production and distribution of goods and services or to generate and manage knowledge and information. Life is cheap, and freedom of choice, expression or movement is at the discretion of government. It is certain that Nigeria cannot handle a serious natural disaster or war with a serious, even if minor, power such as apartheid South Africa. Social interaction is inefficient. Interpersonal communication is diffuse and time consuming. Leisure is either nonexistent or incidental especially among the urban population. Expectations are high as a result of information flow from outside the country but expectation realisation levels are so low that frustration leading to fatalism and anti-social behaviour result.

Many reasons or explanations have been advanced for this state of affairs, the most popular of which are listed in Table 1. In this paper, I shall restrict myself only to issues related to engineering and technology and how they may be used to improve the lot of Nigerians vis-à-vis the world situation.

Table 1: Most Common Reasons Given to Explain Nigeria's Underdevelopment.

Individual Cause	National Cause	International Cause
Tribalism	Bad government /leadership /followership	Machinations of former colonial masters
Nepotism	Federal character or lack of it	Capitalism
Greed and corruption	Tribalism	International racism
Lack of patriotism among Nigerians	Greed and corruption	Reluctance of advanced nations to transfer real technology
Poor work ethics/attitude	Lack of patriotism among Nigerians	

Technology

Although technology may have been defined too frequently, I crave your indulgence to present only two of the more modern and practical definitions which I believe are relevant to this topic. The first definition is that technology is a complex system of interactions between the history, culture, physical and resource environment, social, political and commercial objectives and practices of any society or body, in a process in which that body or society attempts or has attempted to solve its problems or to sustain itself (Nnolim, 1985). In other words, according to Patrick Moynihan, Democratic Senator for New York in the 1970s. *"Technology is not a product, it is a system"*.

The second definition insists that technology is simply the business of providing goods and services to human beings in a convenient, affordable and acceptable manner.

Nigerian Technology

These definitions imply, first of all, that Nigeria, like all other societies, has indigenous technologies which can be definitely called Nigerian. It is these technologies that the areas, which came to be called Nigeria, used, before the white man came, in agriculture to cultivate their crops, and rear their livestock, in building and construction to build their homes and other structures and in medicine, education, etc. to serve their needs in these areas.

Secondly they show that Nigeria must incorporate some version of imported, what I call "world", technologies, if it is to survive. Space and time constraints do not permit elaboration but the following points should be made:

- Nigerian indigenous technology is still largely responsible for meeting the daily needs of the majority of our population even though it can no longer, and has never been able to, cope.
- Indigenous technology is in the hands of people who cannot improve upon it.
- Imported technology is much more capable of coping with current

needs.

- Indigenous expertise in imported technology is limited and largely out of date.

Nigerian Engineering

Whatever we chose to call Nigerian engineering may be better understood if examined through historical perspective. It is necessary to bear in mind that, up to 1989, the popular and official government understanding of what constitutes engineering was limited to civil, structural, mechanical and electrical engineering in that order. This engineering, up to 1940 in Nigeria, was essentially dominated by British professional practice. Its purposes were directed at the execution of projects approved by the Colonial Office. Major engineering projects such as railways, bridges, canals etc, were designed and planned in Britain by British engineers with data obtained largely under the supervision of semi-professionals who worked with local labour.

Indigenous input was largely in terms of manual or semi-skilled labour. With very few exceptions, the highest technical position of the Nigerian was foreman or supervisor. These foremen and supervisors were at best semi-literate but were, more often than not, illiterate and learnt from example. The engineering achievements of the time, therefore, were largely due to the power of British colonialism and the fear of that power by the Nigerian technician.

The second world war of 1939 - 1945 opened the eyes of both the British government and the African to their interdependence and by the, late 1940s there was a dramatic increase in the number of trained Nigerian technicians and engineers.

The thinking of the time favoured the training of more technicians than engineers. The very few Nigerians with engineering degrees lacked the experience, exposure, confidence and the authority of their British counterparts. This situation continued through the late 50s to the 60s even though the few Nigerian engineers there were, had at Independence, taken over from the British.

Up to 1970, except in Biafra during the Civil War, Nigerian engineering made little technical impact on society. It also got involved in practices such as taking kick-backs from contractors and from falsified material supply or purchase transactions. By 1970 it had become clear that Nigerian engineering could not cope. With post war rehabilitation and the oil boom, Nigeria had very little choice, if the services were to be provided, than to import foreign and technical expertise. This must have awakened the government and indigenous engineers to the national security and survival implications of the situation.

The late 1970s and early 1980s saw greater attention being paid to the role of the professional societies and of government, in engineering practice and education. The new universities of technology, polytechnics, the 6-3-3-4 system of education etc, are all products of this era.

Societal Progress and Technology

The progress of nations has been described in one analysis, as the movement from the pre-industrial through the industrial to the post-industrial society (Bell, 1976). Nigeria is essentially a pre-industrial society whose strategic resource is raw materials while its transforming resource is natural power such as human and animal labour. Often the technologies we are willing, or can afford, to import, or develop, belong to the industrial society whose strategic and transforming resources are money capital and created power such as electricity. While we are struggling to come to grips with the demands of this society, the suppliers of the technology or the tools with which we will develop our own, are entering the post-industrial society whose strategic and transforming resources are knowledge and data transmission respectively. All these imply that there is need for a closer look at many of the things we are doing at the national and state levels in government, universities, polytechnics, industry, etc. vis- a-vis the process of development or management of technology.

Models of Technology Management

I shall discuss two models of technology management. In one model, which applies in a corporate context, where specific products and markets can be known and organisational resources are under specific control, there are four phases in technology development (Marvin, 1973):

- The Creation phase when the idea of a product or process is born and defined
- The Co-ordination phase when the methods of realisation or the application of the idea are formulated and planned
- The Commercialisation phase when the prototype of the hardware, systems and associated services are developed and tested
- The Capitalisation phase when money capital is finally committed to the implementation of the idea.

The other model (Nnolim, 1984), which is more individual oriented, defines technology management or development as, merely the solution of the following six problems:

- The Knowledge problem
- The Terminology (jargon) problem
- The Tool Utilisation problem
- The Design problem

- The Operation problem
- The Management problem

Nigerian governments tend to operate directly according to the first model while the society at large operates the second model. For example, all government projects in Nigeria, jump from the creation phase to the capitalisation phase. The co-ordination and commercialisation phases are usually left entirely to foreign expertise and technology transfer agreements and to technical aid packages. The society, at large, concentrates its effort on solving the knowledge, the terminology and to some extent the tool utilisation problems before leap-frogging into the management problem area. The result is, on the government side, involvement in projects in ways that meet the business criteria of foreign industry to the detriment of the Nigerian economy. On the Nigerian peoples' side, the result is poor management and inability to run any enterprise efficiently in a modern sense.

These are not to say that Nigerian governments or people are to blame exclusively especially when viewed against the needs of governments, which have short life-spans, to show some results or those of people under unstable governments to plan for their survival. In several articles, I have apportioned blame for this state of affairs on government policy (Nnolim 1989), work ethics of the average Nigeria (Nnolim, 1987) and the manpower substitution policy of both government and originally foreign owned industrial organisations (Ibid, 1987).

The Problem

The strength of nations rests on many foundations. Depending on what the national objectives are, at any one time, this strength may be connected with one or more of the following:

- natural resources
- human resources
- technology
- geographical and historical fortune.

It is a fact that Nigeria is rich in natural and human resources. Technology of any kind is available to Nigeria. Nigeria has been fortunate, in both geographical location and historical contacts, to be challenged, century after century, by the current of the times. Why then can Nigeria not solve its problems, at least as well as the developed countries? Why can it not use its human resource, available technology, and spatial and temporal location to exploit and develop its resources? My answer is that the problem is the African professional. Please note that by professional, I do not mean professional as used in the government ministries to mean only engineers, doctors etc. but as any person who has received and graduated from specialised training in an endeavour from which he earns his living. A few examples will suffice:

- The average electronic technician or motor mechanic, whether on the roadside or in the bigger companies, regularly, dupes his less informed client or employer by diagnosing non-existent faults in order to make money.
- The average board member, chairman or managing director of a corporation regularly angles to have company resources diverted to his personal use by employing one subterfuge or another.
- All the ministries of works, banks (commercial or otherwise), utility companies, departmental stores, manufacturing companies are inefficient because the professionals there are inefficient. They make many claims to profitability, often authenticated by sales records, but close analysis reveals that inefficient operations are often covered by excessive charges.

The Solution

If the premise is accepted, that the African professional is the problem, then the problem has only one solution namely – improve or upgrade professionalism in Nigeria. This apparently simple solution, however, is not so simple because the question arises: "Who will do the improving or upgrading"?

Having tried to answer the question over the years, I called in one paper (Nnolim, 1989) for a new generation of engineers and technologists. This new generation would:

- have a proper understanding of engineering and technology and not pass off applied science, such as physics, mechanics etc. as engineering or technology
- have social perspective and goals
- have social commitment and responsibility
- avoid fraud and indolence.

It is difficult to get such a new generation unless the educational institutions, training such people, make deliberate efforts to get them. It is near hopeless expecting government or the Nigerian professional societies to do so knowing the Nigerian talent in loop-hole technology. Our best chance (note the word "chance") is for the educational institutions to strike a proper balance, in their training of students, between the phases of corporate technology management or between the six individual technological problem areas. The latter may have been the reason for recent thrust of intensive activity by the National Board for Technical Education (NBTE) in Nigerian polytechnics and may have informed the decision to establish the universities of technology. The question is "will these institutions, their staff and students take the challenge?"

Conclusion

Nigeria is not bereft of its own indigenous, nor of access to imported, technology nor the human and material resources to develop them. Its problems are man made and hinge on its lack of appreciation of, and commitment to, genuine professionalism. Models of technology development, which have proved successful, as well as products of modern technology, present us with yet another chance to catch up with the rest of the world. It is our duty to ensure that we do not miss the opportunity again.

References

1   Nnolim B.N.; (1985) Environmental and Manpower Constraints in the Management of Technology in Nigeria; IMT Convocation Lecture.

2   Bell D.; (1976); Welcome to the post-Industrial Society; Chem Tech. Vol. 10. No. 6 pp 608 - 610 Am. Chem. Soc.; Washington D.C., USA

3   Marvin P; (1973): Fundamentals of Effective R & D Management, ACS Audio Course: Am. Chem. Soc. Washington D.C., USA

4   Nnolim B.N. (1984): Technological Emancipation of Nigeria: A Lecture presented to Final Year HND Mechanical Engineering Students, IMT, Enugu.

5   Nnolim B.N. (1989): Key Chemicals for the Nigerian Economy and Industry: Proceedings of the 1989 AGM and Conference of the Nigerian Society of Chemical Engineers, Lagos.

6   Nnolim B.N. (1987): Improving Work Ethics in Nigeria: Submitted to the Nigerians Institute of Policy and Strategic Studies, Kuru near Jos.

7   Nnolim B.N. (1989): The Call For a New Generation of Engineers and Technologists: A paper presented at the launching of the IMT Student Chapter of the NSChE.

1.3: The Call for A New Generation Of Engineers And Technologists

Being a paper presented at the launching of the I.M.T. student Chapter of the Nigerian Society of Chemical Engineers on 29 April 1989

Ladies and Gentlemen,

It is usually the custom, at gatherings of this nature, for practitioners, under the same discipline, to deliberate, eulogise and very rarely criticise, various aspects of their discipline. The benefit is that at the end of the exercise, each participant would have had the opportunity to compare his or her experience and knowledge against those of others and increase his pride in his profession so that the end result is a general improvement in the level of proficiency in that discipline. While in agreement with these observations, the subject matter of my brief talk today will not be the now clichéd sermonizing on technology, industrialisation, local sourcing etc. I will like to deal with a fundamental issue which is not quite so complimentary.

The Reality

The current situation is that Nigeria is backward country, rich in largely untapped, under-utilised or wasted human and material resources. There is an ineffective indigenous capacity to feed, house, clothe or protect Nigerian citizens or their health. Nigeria has not been able to develop or sustain an adequate means of production and distribution of goods and services or to generate and manage knowledge and information. Life is cheap and freedom of choice, expression or movement is at the discretion of government. It is uncertain that Nigeria would be able to handle a serious natural disaster or war with a serious, even if minor power such as apartheid South Africa.

The Problem

The strength of nations rests on many foundations and depending on the issue, the critical foundation may be connected with one or more of the following:

(a) natural resources
(b) human resources
(c) technology
(d) geographical and historical fortune.

It is a fact of history that both the colonial and indigenous governments of Africa did invest, in spite of what we are being asked to believe by the so-called nationalists and political patriots, in the development of the natural, human and technological resources of their countries. African nations, have been fortunate, in both their geographical location and historical contacts, to be challenged century after century by the current of the times.

The problem of Nigeria, and indeed of Africa, is that, as its human resource becomes more developed, any existing or consequent technology, unique geographical location and historical fortunes are used to exploit African natural and human resources to the detriment of Africa. More specifically, those Africans who rise above others, either in knowledge, wealth or power, regularly, use these advantages, almost every time, to the detriment of Africa and their fellow countrymen.

For example, powerful chiefs, who came into contact with the Europeans, organised and sold off their countrymen as slaves. The average electronic technician or motor mechanic, whether on the roadside or in the bigger companies, regularly dupes his less informed client by diagnosing non-existent faults in order to make money. The average African government minister or president regularly ensures that all public projects executed under his regime do not proceed without his personal benefit either in strict financial remuneration (usually in foreign exchange), in power patronage by him or in the massage of his ego.

My contention here, to cut a long story short, is that the problem with African technology and development is the African professional. Please note that by professional I do not mean professional as used in the government ministries to mean only engineers, doctors, accountants, etc. but as any person who has received and graduated from specialised training in an endeavour from which he earns his living. I have also carefully avoided the use of the word-elite - because it is a word which has lost its meaning in Nigeria and will, therefore, be misunderstood by many.

Detailing the Problem

Because we are Nigerians and I am addressing a largely scientific and engineering oriented audience, I shall concentrate only on Nigeria and on engineering and associated professions. I shall also treat the following periods namely up to 1950, 1950 - 1960, 1960 - 1970, 1970 -1980, 1980 - 1990 and 1990 and after as periods of significance.

Engineering up to 1950 in Nigeria was essentially dominated by British professional practice. The overall in-charge was the colonial administrator who was usually not an engineer. The engineer merely executed projects approved by the Colonial Officer or D.O. The major engineering projects such as railways, bridges, canals etc. were designed and planned in Britain by British engineers. The data input and execution were largely under the supervision of British semi-professionals who worked with the local labour. With the exception of a very small minority of native engineers, the highest technical position of the Nigerian was foreman or supervisor. These foremen and supervisors were at best, semi-literate but usually illiterate and learnt by example. The issues have been glamourised, in the recent past, in nationalist and anti-racist propaganda but the dominant characterisation, at the time, of the Nigerian technical worker, was that he was indolent and dishonest. The only way to get him to work honestly was to browbeat and intimidate him into it and to sack him on the spot for any suspected fraud. The Nigerian response was that it was white mans' work and, therefore, why should he put his heart into it? It is better to work for one's self and one's own country. The engineering achievements of the time, therefore, were largely due to the power of British colonialism and the fear of that power by the Nigerian technician.

The second world war of 1939 - 1945 opened the eyes of both the British government and the African to their interdependence and by the 1950s there was a dramatic increase in the number of trained Nigerian technicians and engineers. The thinking of the time favoured the training of more technicians than engineers and to have a City and Guilds certificate in the 40s and 50s was more in those days than a PhD is today. This group was much more obedient to colonial instruction but generally lacked initiative or the audacity of their less trained predecessors. The few Nigerians, with engineering degrees, lacked the experience, exposure or the colonial authority of their British counterparts and were, therefore, not much respected by the Nigerians technician or rank and file artisan or worker.

At independence in 1960, the few Nigerian engineers there were, had no choice but, and indeed relished the long sought opportunity, to take over from the British colonials. The technical cadre had no choice but to subordinate themselves to this new crop of native engineers for whom they had little respect. Between 1960 and 1970, except in Biafra during the Civil War, there was very little of engineering being done. Most of the period was spent by the engineers in exploring and obtaining the welfare benefits of the white man's posts as they imagined them to be - exercising unquestionable authority over reluctant native workers, garnering every perk of office and deciding, as took their fancy, that sand was laterite or laterite was sand or that the grossly inferior lumber, known locally as "akpu" can serve the same purposes as the highly priced "Iroko" timber etc. The technicians, finding their recently acquired skills not much understood or valued by the new engineers and not being trained for anything else, sought refuge in surviving within the empire of the new masters. All this time, everybody seemed to have forgotten the almost slavish devotion to duty, attention to detail and obsession with results, economy and efficiency with which the colonial engineers and technicians were associated, both in fact and in the eyes of the populace.

By 1970, it had become clear that Nigerian engineering either did not exist or could not tackle the already accumulated and rapidly accumulating problems. With the oil boom, Nigerian really had very little choice, if the services were to be provided, than to import foreign engineering and technical expertise. Since they had also lost their monopoly, of deciding that ten trips of laterite was the same as one hundred, to administrators, who saw very little engineering in falsifying figures for a little bribe, the engineers had no choice than to work up a new scheme of survival. Such schemes included the setting up of professional engineering organisations ostensibly to improve engineering practice although the same administrators, in government, saw it as ploy to improve their contract bidding chances. When this did not quite succeed, Nigerian engineers aligned themselves with foreign engineers and this seemed to have improved their contract bidding prospects more than it did for Nigerian engineering. The trained technician had, of course, very little part to play in all this. Very few new technicians were being trained or wanted to be trained. The older ones were jostling to be treated as engineers and the first school leaving certificate holder was at the roadside taking business from everyone else. Since all this was waste, too much waste for what was being paid for, the bubble had to burst. When it burst around 1979/80, few Nigerians believed it until, around 1983 when, General Buhari and Brigadier Idiagbon, after their successful military overthrow of the Shagari civilian federal government, forced the news down everyone's throat.

The 1980s have seen much more sobriety and humility among all Nigerians. The professional engineering societies have become more serious. More engineers are unemployed and the training and experience of engineers have become issues of concern to government. There is still no identifiable and acceptable Nigerian engineering although there is a lot of challenge and opportunity under these Structural Adjustment Programme (SAP) conditions for Nigerian engineering to come into its own. It is because I do not see Nigerian engineers and technicians taking the challenge in the 1990s that, like many town unions in Anambra state, and now the Federal government, have called for a new generation of politicians, I am calling for a new generation of engineers and technicians. We cannot, as in politics, ban the old and existing engineers. But we can plan for their extinction by not swelling their ranks and by training new engineers and technologists who will not operate like the old. How can this be done?

The New Engineers and Technologists

The first step is to ensure that what was wrong with the old engineers and technologists do not repeat themselves in the new generation. These are:

- Lack of Understanding of Engineering And Technology Itself.
  Much of "brilliant" Nigerian engineering is often applied science such as physics, mechanics and such profit, cost or schedule unconscious disciplines. Engineering, on the other hand, is getting things done as designed, on schedule and within acceptable cost limits for human benefit.

- Lack of Social Perspective.
  Nigerian engineering has no vision of the Nigerian society it is building. All that concerns it is the bridge or road at hand. What this bridge or road does to society, economy or to individuals is not its concern. This, of course, is not engineering.

- Lack of commitment and social responsibility
  Much of real engineering is concerned, to borrow from Bishop M. Eneje of the Enugu Catholic diocese, with the quest and ability of man to "master his circumstance". When there is money, Nigerian engineering will squander it in opinion, rather than, empirically determined systems. When there is no money, Nigerian engineering will demand a bill that is twice what there is for the entire business of the populace and will do nothing until it is provided. Nigerian engineering is a practice of excuses - how the roads cannot be built because of the rains, how electric light cannot be supplied because there are no spare parts, how there can be no water at X location because it is uphill from Y location etc. These are not engineering but social irresponsibility and lack of commitment which result in submission to the tyranny of

circumstance.

- <u>Fraud and indolence</u>
    It is a philosopher's dilemma and a manager's nightmare to have to chose between a brilliant but fraudulent and indolent worker and an industrious, committed but not so bright a worker. Nigerian engineers are extremely bright and, in a class room situation, may rank among the top five in the world. In practice, however, the mix of indolence and dishonesty (material and intellectual) seem to take over except under strict expatriate supervision.

The second step is to expose students and practitioners to engineering as it is practised all over the world. This is not necessarily by paid trips to other countries of the world but by making it possible for modern engineering aids such as computers and machines as well as engineering literature to be accessible to Nigerian engineers and students. Here, I have in mind tax deductible subscriptions to international professional engineering associations and journals, tax deductible purchase of computers by individual engineers and institutions, an increased availability of UNESCO coupons to engineers and students.

I believe that there are some who may not accept this analysis of Nigerian engineering or the recommended solutions. I hereby and graciously grant them, their point of view. Ours is an empirical business. Truth in our profession does not depend on your authority or position or wealth or on whom you are associated with. It is from our fruits that we are known.

Whatever our view point, our reasons, or explanations, my position is that Nigerian engineering has not worked and needs to work. I believe that if we, as engineers, endeavour to really understand engineering, take more seriously the non-engineering but social, ethical and moral issues of our time, determine to be more committed, more socially responsive and responsible, with a little more integrity, the future is bright.

Thank you.

CHAPTER TWO

THE ROLE OF CHEMICAL ENGINEERING

2.1: The Chemical Engineer And Society
An Orientation Lecture for First Year Chemical Engineering Students

Human society can be regarded as a system, controlled, by humans and having certain distinguishing characteristics, objectives and practices. The aspect of society that concerns us today is its WELFARE and this I choose to see, for our purposes, in terms of Abraham Maslov's hierarchy of needs although it is not the only view. These needs are, starting from the lowest,

- Survival Needs
- Security Needs
- Belonging Needs
- Status Needs
- Self Actualisation Needs

The major pre occupation of human beings, therefore, seems to be the satisfaction of these needs. Recorded history, though not complete, illustrates the shift in emphasis from survival to self actualisation needs, depending on the sophistication of the society.

In fact, the original motivation to mechanisation of tasks was to transfer the menial tasks necessary for survival and security (physiological or animal needs) to machines in order to give man more time to pursue human or psychological (belonging, status and self actualisation) needs. This search, together with greater understanding of the physical world, economic, political and other developments, eventually culminated in the industrial revolution of the 17th and 18th Centuries.

Engineering, on its own, is probably older than recorded history. The first practitioners recognised as engineers, however, were the Egyptian architects who are credited with the erection of the pyramids etc. Then came the Roman engineers who built the Roman roads and aqueducts throughout the Roman Empire. Many other engineering feats and wonders have been recorded since then. At no time, however, has the scale and pervasiveness of the ability to satisfy mans' material needs, as defined by Maslov and others, reached the level it has in the last one hundred years. Although this period is characterised by rapid and profound developments in all fields of human endeavour in the arts, sciences or engineering, it is the contention of this paper that chemical engineering was the key to the scale and cohesion of these developments.

This is because it was, and still is, the only discipline that is concerned with changing the properties of material in bulk. When Maslov's hierarchy of needs are translated into common parlance they reduce to the basic needs for food, clothing and shelter, health and hygiene, transport and communication, power and energy, culture, education and leisure. It is the chemical and allied industries which provide for these needs. And it is chemical engineers who run, or are instrumental to the running of, these industries.

In the clothing and shelter area, they develop and produce, working with chemists, metallurgists and textile technologists, raw materials for foods, equipment, housing and clothing such as steels, aluminium, and other metals, cement, paint, glues and other building materials, plastics, synthetic fibres and dyes etc.

In the health area, they work with pharmacists and medical personnel to manufacture various disinfectants, fungicides, bactericides and numerous drugs. Chemical engineering science is the key to the development and design of the artificial kidney, the nuclear pacemaker, the maintenance of sterile environments and prostheses made of carbon. In occupational hygiene, chemical engineers are involved in pollution and noise control and in waste water and sewage treatment.

In the transport, power and energy area, chemical engineers have extended fuel and power supply and utilisation flexibility by providing them in different forms such as petrol, diesel, fuel oil, town gas, aviation fuel, LPG etc. They are in the forefront of fuel cell development, solar energy, power plants, thermal energy stations etc. They are responsible in nuclear energy work for processing of radioactive materials and wastes.

In the communication and education area, chemical engineers work with materials scientists and physicists to produce metal cables and optical fibres for information transmission, solid state wafers, plastics and photo resists for the production of memory devices for computers. With paper technologists they engineer and operate processes for paper manufacture.

The impact of chemical engineering on plastics and other materials is, in fact, the main support on which the rapid expansion and sophistication of the entertainment, image making and public enlightenment industries are based.

It is clear, therefore, that chemical engineering and the chemical engineer are crucial components in our developing industrial society. This imposes very strict constraints on those of us who have chosen the profession in Nigeria.

First we cannot afford, being a socially and militarily strategic profession, to join the cacophony of professional voices jostling for the attention of a seemingly unconcerned public. Yet the strategic position of our profession dictates that we must have their attention.

Secondly we cannot afford, in the practice of our profession, to join the ranks of other engineers bemoaning the absence of this or that incentive or funds even though our profession is a capital intensive one.

Thirdly since chemical engineers are the most broadly trained of all engineers, learning relevant aspects of mechanical, electrical, civil, and structural engineering in addition to chemistry, material science, mathematics and management sciences, the onus is on chemical engineers to provide leadership by example in the manner in which they communicate and associate with other professionals in the solution of Nigeria's problems.

These are the benefits and the tasks ahead of you as chemical engineering students. The measure of success or failure you have as a chemical engineer will depend on how far you have endeavoured to benefit from the near excellent training which is offered in this Department.

Thank you.

2.2: Nigerian Technology and the Use of Local Raw Materials - The Role of Chemical Engineering

Presented at the Anambra State Science and Technology Week, 16 - 21 October 1989; Enugu

Introduction

Local sourcing of raw materials became fashionable recently when the foreign exchange demands of the import substitution industries could no longer be met by our export earnings. Sadly too, these industries were largely peripheral and parasitic to the industrial development of the country. The benefit of the situation, however, is that Nigeria has been forced to look inwards and to seek to apply its own resources to the solution of its problems. How we do that and tackle the problems can make all the difference. This paper is an attempt, from one perspective, to analyse the issues involved. This perspective is grounded in the belief that technology applied to the processing of raw materials can result in economic and industrial progress of nations.

Technology

Although technology may have been defined too frequently I crave your indulgence to present only two of the more modern and practical definitions which I believe are relevant to this topic. The first definition is that technology is a complex system, of interactions between the history, culture, physical and resource environment, social, political and commercial objectives and practices of any society or body, in a process in which that body or society attempts or has attempted to solve its problems or to sustain itself (Nnolim; 1985). In other words, according to Patrick Moynihan, Democratic Senator for New York in the 1970s, "Technology is not a product; it is a system". The second definition insists that technology is simply the business of providing goods and services to human beings in a convenient, affordable and acceptable manner.

Nigerian Technology

These definitions imply, first of all, that Nigeria, like all other societies, has an indigenous technology which can be called Nigerian. It is this technology that the areas now called Nigeria used, before the white man came, in agriculture to cultivate their crops and rear their livestock, in building and construction to build their homes and other structures, and in medicine, education, etc. to serve their needs in these areas.

Secondly they show that Nigerian technology must incorporate some version of imported technologies if it is to survive. Space and time constraints do not permit elaboration but the following points should be made:-

- Nigerian indigenous technology is still largely responsible for meeting the needs of the majority of our population even though it can no longer cope,
- Indigenous technology is in the hands of people who cannot improve upon it.
- Imported technology is much more capable of coping with current needs.
- Indigenous expertise in imported technology is limited.

Raw Materials

Raw Materials are those materials which are in their natural or untreated state. In Nigeria, raw materials have come to be understood as anything that is used to make a product. This seemingly trivial point has been a hindrance to raw material development, industrial capacity utilisation and the economic progress of the Nigerian State.

It sounds like mere semantics but the truth is that, with very few exceptions, industrial products or consumer goods are not made DIRECTLY from raw materials. These products are made from INTERMEDIATES which can be primary, secondary, tertiary etc. derivatives of raw materials. Thus, to take a very simple example, iron ore is a raw material which can be processed to ingots (or billets), a primary intermediate. The billet can be cast or rolled to any shape such as a rod or flat bar which becomes a secondary intermediate. This rod or bar can form a final product such as a window protector or can become a tertiary intermediate, such as a bolt, in another product and so on. The point is that our understanding of raw materials and our knowledge of their intermediates determines how and to what level we source and use such materials. Such use normally results in the growth of several industries preoccupied with either direct raw material processing to intermediates or the use of those intermediates for the manufacture of industrial or consumer products.

The second point to clear is what we regard as local when discussing raw materials. I get the feeling, from popular media usage and Government statements, that crude oil, natural gas, tin ore, columbite and the like, found in Nigeria, are not local raw materials and that cassava, palm oil etc are so peculiarly Nigerian that they cannot be found anywhere else. This feeling is reinforced by the fact that crude oil, natural gas etc. are regarded by government solely as "international raw materials" which may be grudgingly eked out in Nigeria only when absolutely necessary in .order to stem the drain on foreign exchange caused by the importation of their derivatives such as petrol, carbon black etc.

In the same vein, there is a lot of to do about research in local raw materials outside the oil industry as if these materials exist only in Nigeria and only since 1960. The truth is that any material found in Nigeria is local to Nigeria and that, except for a few exotic jungle herbs, most raw materials and intermediates of industrial or any importance are not only known throughout the world but have a manufacturing process. To exclude the "international raw materials" from local and general development and to embark on the development of the so called local raw materials as if they are new or hitherto unknown materials is therefore ill-advised. All modern societies require essentially the same mix of goods and services for which the raw materials, intermediates and technologies are well known and developed. Our task is to associate those materials found in our localities with their intermediates, products and technologies and exploit them for our benefit. This is not to say that certain raw materials, intermediates and products, especially in food, are not peculiarly Nigerian and therefore do not need further developmental effort. Whether the task is that of the identification, acquisition and adaptation of existing technology or that of developing new technologies, the process and context of the development of technology need to be properly understood.

The Context

Nigeria is essentially a pre-industrial society (David, 1982) whose strategic resource is raw materials while its transforming resource is natural power such as human and animal labour. The technologies we may wish to acquire or develop belong to the industrial society who's strategic and transforming resources are money capital and created power such as electricity etc. which are in short supply here. While we are struggling to cope with the demands of this society, the suppliers of the technology, or the tools with which we will develop our own are entering the post - industrial society who's strategic and transforming resources are knowledge and data transmission respectively.

The Process

The process of technology development in the context of the present paper can be illustrated by two models. In one model, which applies in a corporate context, where specific products and markets are known and organisational resources are under specific control, there are four phases in technology development (Marvin, 1973).

- The Creation phase when the idea is born and defined
- The Co-ordination phase when the methods of realisation or the application of the idea are formulated and planned
- The Commercialisation phase when the prototype of the hardware, systems and associated services are developed and tested
- The Capitalisation phase when funds are applied to achieve the objectives of the project

The other model (Nnolim, 1984), which is more individual oriented, defines technology management or development as, merely, the solution of the following six problems:

- The Knowledge problem
- The Terminology problem
- The Tool utilisation problem
- The Design problem
- The Operating problem
- The Management problem

Nigerian governments have tended to operate directly according to the first model while the society, at large, has operated the second model. These governments, in government sponsored projects, tend to jump from the creation phase to the capitalisation phase often leaving the co-ordination and commercialisation phases to foreign expertise and to technology transfer agreements and technical aid packages.

The society at large often struggles to solve the knowledge, and to some extent, the terminology problem and then leap frogs into the management problem area: The result is, on the government side, involvement in projects that meet the business criteria of foreign experts to the detriment of the Nigerian economy. On the Nigerian peoples' side, the result is poor management and inability to run any enterprise efficiently even if given on a platter of gold.

The implication for Nigerian technology and the use of local raw materials is obvious:
- train Nigerians and get Nigerians already trained, in the co-ordination and commercialisation phase, or in the tool utilisation, design and operating problem, activities.
- reorient Government policy to encourage indigenous activity in all the phases of technology management especially those areas forming part of most technology transfer agreements

The Role of Chemical Engineering

Modern technology is the product of efforts of teams of professionals and individuals and it would be erroneous, at any time, to ascribe to any individual or profession sole wisdom and credit in any technological development. Nevertheless, certain professions constitute change agents and play the most prominent and significant role in certain fields of endeavour.

In the area of industrial development, where it is concerned with the provision of the essential necessities of life and the enhancement of the general well being of the people through provision of vital human requirements in food, clothing, shelter, health and hygiene, transport and communication, culture, education and leisure, power and energy, chemical engineering has been the change agent in this century (which has seen the greatest development throughout human history) (Onwuka & Nnolim, 1981).

Chemical engineering can do this because it is the only branch of engineering concerned with processes in which materials undergo a required change in composition, energy content or physical state, with the conception, development, design, improvement and application of processes and their products; the economic development, design, construction, operation, control, and management of plant for these processes and with research and education in these fields (I. Chem E, 1977).

Compare the training of chemical engineers with the processes for the development of technology and it becomes clear that chemical engineering has to LEAD other professions in the joint effort to develop and use our raw materials. It is known from experience that the needs of most modern societies are served, from the chemical engineer's point of view, by the following groups of chemicals shown in Table 1. The development of these resources, most of which are present in Nigeria, is stultified by the inability of government to involve the generality of Nigerians, through privatisation, in their exploitation.

The current efforts at local sourcing have been concentrated at developing local inputs for peripheral industries such as breweries, bakeries etc. This will save some foreign exchange but cannot do much more. The current NDE (National Directorate of Employment) and DFRRI (Directorate of Food, Roads and Rural Infrastructure) projects are valuable as sensitisation projects to alert the nation on the need for self reliance and employment generation at the grassroots level. To, really, address the national problems which they seek to solve and achieve multiplier effects of their activities, greater emphasis needs to be made on getting more technical and professional support and orientation into them. There is need to strike at the heart of critical industry such as the indigenous food industry, the chemical industry and the metallurgical industry. These are areas in which chemical engineers and technologists have played leading roles in the developed world.

The Food Industry in Nigeria consists of, on one hand, the organised industry producing foreign beverages, snacks and cereals and, on the other hand, the unorganised sector providing the bulk of indigenous food and condiments. There is need to apply modern technology to the indigenous food chain preferably through sponsorship of serious research and development efforts at the local universities, polytechnics and research institutes. This will yield more fruitful dividends than the search for local substitutes of imported beverages, snacks and cereal products which can be left to the industries concerned.

The metallurgical industry should be decoupled from theoretical economic postulations and set down squarely on the plains of fact and common sense. Ajaokuta, Oshobo, Aladja, Jos and Katsina steel or rolling mills should be privatised if possible or scaled down to capacities commensurate to the needs of the country.

Table 1: Key Chemicals in a Modern Economy (7)

S/No	Group Name	Key Chemicals	Major End Uses
1	Oil based Petrochemicals	Ethylene, Butadiene, Propylene, Benzene, p-Xylene	Fabricated Plastics, Antifreeze, Resins, Fibers, Elastomers
2	Natural Gas based Petrochemicals	Ammonia, Urea, Methanol, Formaldehyde	Fertilizers, Explosives, Plastics & Fibers, Adhesives, Animal Feeds
3	Chlor-Alkalis	Chlorine, Caustic Soda, Soda Ash	Organic Chemicals & Plastics, Pulp & Paper, Chem. Mfg., Soaps and Detergents, Glass
4	Acids	Sulphuric Acid Phosphoric Acid	Fertilizers, Metal Processing, Animal Feeds
5	Industrial Gases	Oxygen, Nitrogen, Carbon Dioxide, Hydrogen	Metal & Chemical Mfg, Refrigeration, Electronics, Petroleum refining
6	Mineral based Inorganics	Lime, Sulphur, Potash, Phosphorus	Metallurgy, Chem. Mfg, Fertilizers, Detergents, Foods, Water Treatment
7(a)	Plastic Monomers	Styrene, Vinyl Chloride, Propylene Oxide	Fabricated Plastics & Coatings
7(b)	Plastic Polymers	Phenolics, Polyesters, Epoxies	Adhesives, Engineering Plastics, Coatings & Laminates
8(a)	Fiber Monomers	Dimethyl Phthalate/Purified Terephthalic Acid (DMT/PTA), Ethylene Oxide, Cyclohexane	Fibers & Films, Surfactants, Solvents
8(b)	Fiber Polymers	Polyester, Nylon, Acrylics	Apparel, Home Furnishing, Industrial
9	Adhesives & Coatings	Phenol, Vinyl Acetate	Adhesives, Plastics, Paints, Coatings
10	Pigments	Carbon Black, Titanium Dioxide	Tyres, Rubber products, inks, paints
11	Solvents	Ethanol, Acetone, Methylene Chloride	Laundry, Degreasing

Conclusions and Recommendations

We have identified that:-
- Good technology has to provide goods and services in a convenient, affordable and acceptable manner.
- Nigerian technology exists but cannot cope unless it incorporates foreign technology.
- The new Nigerian technology must be involved in solving the tool utilisation, design, operating and management problems of our society.
- Raw material development policy must emphasise processing to intermediates as this leads to greater industrial activity than the processing of raw materials dedicated to limited and specific end products.
- Raw material development policy must recognise all raw materials found in Nigeria as local and, therefore, not exclude some of them from general attention.
- Raw material development practice must search and obtain licenses for known processes.
- For materials with or without known processes, development or adaptation must use the expertise of professionals such as chemical engineers.
- Emphasis on raw material development should shift from peripheral industries such as breweries and bakeries to the indigenous food chain, the chemical industry and the metallurgical industry.

References

1. Nnolim B.N. (1985) Environmental and Manpower Constraints in the Management of Technology in Nigeria; IMT Convocation Lecture 1985.
2. David E.E. (1982): The Effect of Society on Technology; Chem Tech Vol 12, No.1 pp 19 - 21, American Chemical Society Washington DC.
3. Marvin P (1973): Fundamentals of Effective R & D Management ACS Audio Course, Am. Chem. Soc. Washington D.C
4. Nnolim B.N. (1984): Technological Emancipation of Nigeria. A Lecture presented to Final Year HND Mechanical Engineering Students, IMT Enugu.
5. Onwuka N.D, B.N. Nnolim (1981): Chemical Engineering Key to Industrial Development in Nigeria, official publication of the Nigerian Society of Chemical Engineers
6. The Institution of Chemical Engineers (1977): Rugby UK: Regulations for Election or Transfer within the Institution of Chemical Engineers.
7. Nnolim B.N. (1990) Key Chemicals for the Nigerian Economy and Industry; Proc. AGM, NSChE, Lagos, Nigeria.

2.3: The Place of Chemical Engineering in Raw Material Processing

A paper presented at symposium organised by the Enugu State University of Technology student chapter of NSChE on Monday 22 June, 1992

Abstract

This paper discusses the nature and diversity of raw materials in the context of current concepts of technological societies. The kinds of processing required of raw materials are then discussed in comparison to those techniques and methodologies for which the chemical engineering profession is most expert in. All these are then examined to develop a perspective on the place of chemical engineering in raw material processing.

Introduction

The technological progress of nations has been described, in one analysis, as the movement from the PRE-INDUSTRIAL, through the INDUSTRIAL to the POST-INDUSTRIAL, society (Bell, 1976). These societies are characterised by their strategic and transforming resources.

The pre-industrial society has raw materials as its strategic resource and natural power (such as wind, water, brute animal and human force) as its transforming resource. For the industrial society, the strategic resource is money capital while the transforming resource, is artificial or created power or energy such as electricity. In the post-industrial society; also known as the information society, knowledge is the strategic resource and data transmission its transforming resource. While no present society is 100% pre-industrial, industrial or post-industrial, the societies of Western Europe and North America are more or less post-industrial societies while third world countries are in various stages of transition from the pre-industrial to the industrial society. This situation is further complicated by the inevitable and historical associations and interactions between these different societies. It is, therefore, true that raw material processing technology of those societies that have the most experience in it will be the dominant technology. These are the technologies discussed in this paper.

Raw Materials

A raw material is any substance which is in its natural, unprepared or untreated state. Only raw materials of industrial importance will be discussed. These are (i) mineral and inorganic raw materials (ii) petroleum/organic raw materials and (iii) agro-based raw materials. They are listed in Table 1. A raw material may also be defined, in more general and practical terms, as any substance with is capable of being upgraded to more valuable product in an industrial value addition process. Thus if iron ore is a raw material to the steel industry according to the first definition, ethylene, is also a raw material to the chemical industry, according to the second definition. In the latter case, the raw material may also be regarded as an intermediate. Table 2 lists some of the more important groups of chemical intermediates.

Table 1: Primary Raw Materials

Minerals	**Inorganics**	**Petroleum/Organic**	**Agro-based**
Chromium	Limestone	Crude Oil	Sugar
Alumina/Bauxite	Gypsum	Natural Gas	Maize
Cobalt	Rock Salt	Oil Shale	Cassava
Zinc	Asbestos	Tar Sands	Sorghum
Iron		Coal	Biomass
Copper			

Raw Material Processing

Let us now look at the types of processing required to upgrade raw materials to useful products.

Minerals/Inorganics

Most mineral raw materials occur as solids and as ores or

mixtures of ores and silicates in an earth matrix The major exceptions are dissolved or trace components of liquids, solids or gases (such as salt in sea water, sulphur in coal, in crude petroleum or in natural gas) or homogenous belts and sediments in the earth's crust (such as a seam of coal, rock salt, limestone, gypsum etc). Sometimes, solid minerals occur as nodules on sea beds (manganese) or in pockets in the earth (gold, diamond).

Processing required for these minerals may be grouped under (i) mining and size reduction (ii) beneficiation or primary processing (iii) downstream or industrial processing.

Mining is a unique and distinct discipline which utilises techniques in geology, geophysics, geochemistry, chemistry, civil, mechanical, structural, electrical and chemical engineering to extract viable deposits of raw materials for further processing. Processing here is mainly extraction and separation of materials from earth matter and reducing its size in preparation for beneficiation.

In beneficiation proper, the aim is to increase as much as possible, the concentration or percentage of useful material in the matrix. Processing consists of further size reduction and wet or dry separations. For soluble materials, dissolution in suitable solvent, filtration, evaporation and crystallisation, etc, are also used.

Downstream processing involves the whole range of chemical reactions, physical processes etc to obtain desired products. Details of these are listed in Table 3.

Petroleum/Organic Materials

Crude oil occurs in liquid form and consists of complex and tightly bound organic molecules. Processing of crudes to obtain products, therefore, tends to be severe. The processes are listed in Table 3 also. Typical products from such processing are fuel gas, LPG, motor fuel, kerosene, diesel fuel, fuel oils, lube oil base, asphalt, coke and wax. (Hahn, 1970).

Processing of natural gas is less severe and is directed at producing a gaseous or liquid stream containing only ethane and methane. Severe processing is used if some intermediates are desired as products. Thus removal of sulphur or sulphur compounds, liquefaction, reforming are the more important processes. These are also listed in Table 3.

For Tar Sands, the objective is to separate the bitumen from the sands and thereafter process it by means of conventional petroleum refining techniques. The primary separation method is by hot water flotation. Other techniques include in-situ combustion and steam injection. Oil shale is similarly recovered by the use of above ground hot water retorting, although in-situ retorting is being investigated to save water. Secondary processing of the oil shale to fuels follows after refining. Direct conversion of the main organic component, kerogen, to chemicals is also used (Krieger and Worthy, 1978).

Although coal can be processed as a solid mineral for fuel uses, the processing of coal as an organic raw material is also of industrial importance. Three major types of process are used namely:

Table 2: Main Groups of Chemical Intermediates (Nnolim, 1990)

	Group	Key Chemicals
1	Oil-based Petrochemicals	Ethylene, Butadiene, Propylene, Benzene, p-Xylene
2	Natural Gas-based Petrochemicals	Ammonia, Urea, Methanol, Formaldehyde.
3	Chlor-Alkalis	Chlorine, Caustic Soda, Soda Ash.
4	Acids	Sulphuric Acid, Phosphoric Acid.
5	Industrial Gases	Oxygen, Nitrogen, Carbon Dioxide. Hydrogen
6	Mineral - based Inorganics	Lime, Sulphur, Potash, Phosphorus
7a	Plastic Monomers	Styrene, Vinyl Chloride, Propylene
7b	Plastic Polymers	Polyesters, Phenolics, Epoxies
8a	Fiber Monomers	Dimethyl Phthalate plus purified Terephthalic Acid (DMT/PTA), Ethylene Oxide, Ethylene Oxide, Cyclohexane
9	Adhesives and Coatings Monomers	Phenol, Vinyl Acetate, Ethanol, Acetone.
10	Pigments	Carbon Black, Titanium Dioxide
11	Solvents	Ethanol, Acetone, Methylene Chloride.

- Gasification to synthesis gas which lead to products such as Ammonia, methanol, formaldehyde etc.

- Coal liquefaction to produce coal liquids which are expected to compete with petroleum liquids as fuel or feed stocks.

- Pyrolysis or carbonisation to coke and coal chemicals. Other forms of pyrolysis include high intensity or plasma pyrolysis which yields acetylene and hydro-pyrolysis whose products are methane, aromatics, tar and coke.

Agro - Industrial Raw Materials

The rapid depletion of mineral and petroleum based raw materials in the last one hundred years has focussed attention on renewable resources such as those of animal and plant origin. Agricultural raw materials which were discarded on discovery of synthetic materials are being re-examined for new uses. Such materials are sugar, various starchy grains .and tubers (maize and cassava) and biomass. The major processing method is fermentation usually to ethanol. Other processing methods involve pyrolysis, hydrolysis and anaerobic digestion to liquid, solid and gaseous fuels.

Industrial Chemical Engineering Processes

With the kinds of processing required for raw materials and summarised in Table 3 as background, let us now look at chemical engineering processes. Shreve and Brink (1984) and Perry and Green (1986) have summarised chemical engineering processes as consisting of those based on

1. physical principles of chemical engineering including unit operations, and
2. chemical conversions of chemical engineering otherwise known as unit processes.

These are elaborated in somewhat more detail in Tables 4 and 5. It is clear from these tables, that the methods of chemical engineering are more than suited to the requirements of raw material processing whether it is primary, secondary or downstream processing.

Table 3: Summary of Processing Methods

Mineral/Inorganic	Petroleum/Organic	Fermentation
Mining	Atmospheric Distillation	Fermentation
Beneficiation - Size Reduction - Screening - Classification - DenseMediaSeparation	Vacuum Distillation	Anaerobic Digestion
Tabling	Hydrocracking	Pyrolysis
Jigging	Catalytic Deasphalting	Hydrolysis
Magnetic & Electrostatic Separation	Fluid Catalytic Cracking	
Flotation	Steam Reforming	
Dissolution	Pyrolysis	
Evaporation	Partial Oxidation	
	Hot Water flotation	
	Hot water Retorting	
	Liquefaction	

So what?

Having established that many processing methods for raw material up grades are in fact chemical engineering operations, what next? Is raw material processing simply another name for chemical engineering or vice versa? Do you graduate in chemical engineering and walk right into the coal industry, the sugar industry or the steel industry? Unfortunately no and for a good reason.

In the first place, each of these industries is a specialised industry. Often, the chief practitioners in the industry regard themselves, not as chemical engineers but as specialists in their own right. They would be sugar technologist, coal scientist; metallurgist etc rather than chemical engineers.

Secondly these industries, with one or two exceptions, are older than chemical engineering itself. Thus some of their methods and jargons predate chemical engineering analysis. Old habits die hard and no matter how superior modern chemical engineering analysis may be, good chemical engineers, when they enter these industries or interact with them, make haste slowly, learning the "new" old jargons in the process - and substituting their better methods when they are likely to be most effective and well received.

Thirdly, chemical engineering as a discipline cannot be everything to all things. In spite of its versatility it does not cover every subject and its science must, by the second law of thermodynamics, be deficient in some respects. It is in these respects that chemical engineering must learn from other disciplines especially those that are either related to it or off-shoots.

In summary, therefore, the place of chemical engineering in raw material processing is that of the new kid on the block who has made it but who must listen and learn from the old hands and neighbours that all of them live on the same street.

Thank you

Table 4: Physical Principles and Unit Operations of Chemical Engineering (Perry and Green, 1984)

Principles	Unit Operations
Fluid and Particle Mechanics	Distillation
Transport and Storage of Fluids	Gas Absorption
Handling of Bulk and Packaged Solids	Liquid Extraction
Size Reduction and Enlargement	Adsorption and Ion Exchange.
Heat Generation and Transport	Miscellaneous separation processes
Heat Transmission	Liquid-Gas processes
Heat Transfer Equipment	Gas-Gas, Liquid-Liquid and Solid-solid processing
Psychrometry, Evaporative Cooling, Airconditioning and Refrigeration	Gas - Solids processing

Table 5: Chemical Conversion Principles and Processes of Chemical Engineering (Shreve and Brink, (1984).

Chemical Principle	Chemical Process
Acylation Alcoholysis Alkylation Amination Amonolysis Aromatization	Calcination Causticization Combustion Controlled Oxidation
Carboxylation Condensation	Dehydration Double Decomposition
Dehydrogenation and Hydrogenolysis Diazotisation	Electrolysis
Esterification	Hydrolysis and Hydration
Friedel - Grafts	Ion Exchange Isomerisation
Halogenation Hydrogenation	Neutralisation
Nitration	Polymerisation Pyrolysis or Cracking
Sulphonation	Reduction
	Silicate Formation

References

1. Bell, D (1976); Welcome to the post-Industrial Society, Chem Tech. Vol. 10 No. 6, pp 608 - 610, Am. Chem. Soc. USA.
2. Hahn A.V.G; The Petrochemical Industry - Market and Economics McGraw-Hill Book Co. NY USA, 1970.
3. Krieger J.H and W. Worthy (1978); CHEMRAWN I faces up to Raw Materials Future; Chem. and Eng. News Vol. 56, No. 30. pp 28 - 31; Am. Chem. Soc., Wash D.C. USA.
4. Perry R; Green D; Editors; Chem. Engrs. Handbook 6th Edition, McGraw-Hill Book Co. N.Y. U.S.A. 1984.
5. Nnolim B.N. (1990); Key Chemicals for the Nigerian Economy and Industry; Proc. AGM, NSChE, Lagos.
6. Shreve, R.N. and Brink J.A. Jnr; Chemical Process Industries, 4th Edition; McGraw Hill Book Co. N.Y. USA 1984.

2.4: Chemical Engineering Processes And Pollution

Being a paper presented during the Student Week of the Department of Chemical Engineering, Anambra State University of Science and Technology, Enugu on July, 3 1991.

Introduction

I once read an advertisement, in the Fortune Magazine in the USA, by a small-car manufacturing company. The advert showed two similar photographs of a little man in a big sedan. One of the pictures was a close up of the little man with his head held up high and looking very pleased with himself. The other picture was more panoramic and showed how little the man was in size compared to the car and how wasteful it all looked with all the unoccupied seats and sheer mass of the car. The caption to the picture went something like this "yesterday's symbols of affluence are today's signs of profligacy". Nowhere is this kind of statement tellingly true as in the chemical industry and pollution.

Up to the 1960s, industrialised towns boasted of the number of smoke stacks they had and cities not covered up in smoke were not industrialised. It is only now that we know that pollution, in the final analysis, is really waste - waste of material, human, financial, social, cultural resources.

Very few industries can compete with the chemical and allied industries in the potential and real ability to make such waste. It is, therefore, a very appropriate exercise to examine the role of chemical engineering processes in pollution. Because a number of speakers have been scheduled after me to talk on the subject and I do not know the areas they will treat in their paper, I have decided, with your permission, to gloss over a few general issues.

The issues I shall address are:-
- Pollution.
- Its effects on human life and environment.
- The capacity of the chemical industry to pollute.
- Pollution control measures possible or available.

Pollution

Pollution, expressed in simplest terms, is the contamination of a medium or environment with harmful or undesirable material known as pollutant. The pollution under discussion is environmental pollution where the environment can be air, sea or land. The pollutants are, in many cases, chemical substances but some are inert substances which catalyse undesirable reactions on account of high surface to volume ratio.

Air pollution

There are thousands of air pollutants but the best known are CO_2, SO_X and NO_X gases, soot arising from combustion processes and the chlorofluoro carbons (CFCs). CO_2 concentrations other than natural (0.03%) pose the danger of global warming (the green house effect). Records indicate that CO_2 levels in the air have been around 280 ppm from the end of the most recent ice age. (about 10,000 year ago) until the 19thcentury. Natural cycles of CO_2 (photosynthesis, respiration of plants etc) have fluctuated between 200 - 300 ppm. Since 1958, however, when global measurements of CO_2 became possible, concentrations have increased from 315 ppm in 1958 to 345 ppm in 1986 or an increase of about 13% per year. Other gases such as O_3, H_2O, CH_4 and the CFC,, which also contribute to global warming, show similar trends.

Though the subject is rather more complicated than can be treated here, the effects of global warming are:-
- melting of the polar caps leading to another global flood or ice age.
- changing earthly climate adversely
- worsening the water resource problem by not enabling the recharging of the aquifers (ground water sources)

The SOx and NOx gases give rise to acid gases which cause corrosion of man - made structures and machines and acid rain which is bad for crops. They have also been implicated in some forms of human cancer. The CFCs play active roles in the depletion of the ozone layer which stretches about 60km above the earth, from the troposphere (0 - 15km) through the stratosphere (15 - 42km) into the mesosphere (42 - 82km). CFC-11 ($CFCl_3$), CFC-12 (CF_2Cl_2) have been implicated as well as CCl_4. The role of the ozone layer is to absorb UV and to act as a buffer for O_2 concentration in our atmosphere. When the ozone layer is depleted, skin cancer, eye damage and inhibition of photosynthesis in some species of plants occur.

Land Pollution

The concern here is largely with:
- Contamination of water resources
- Interruption or Destruction of life cycles of plant, animals and micro organisms in the soil
- Excessive loading of cancer causing or debilitating substances in humans.

There are thousands of these pollutants and-only a few will be mentioned here such as 2,4,6 trichlorophenol, 1,2 dichlorobenzene, trichloroethylene, cadmium, lead, cyanide etc.

Sea Pollution

The concern is with aquatic life and contamination of water resources. The same substances involved in land pollution are also involved.

Noise Pollution

Here noise levels above those tolerable to humans or that actually damage human hearing are involved. The noise could arise from operation of machines in the factory, office or from a combination of noises.

Chemical Processes and Pollution

Every chemical process is a potential pollutant. In a refinery plant, for example, sulphur dioxide emissions are obtained from various heaters, furnaces and boilers and have to be treated by amines or by caustic scrubbing. Hydrogen sulphide is removed in the sulphur recovery units. The fluid catalytic cracking units give SOx emissions which have to be treated. In metal coating and treating, chemical processing, and petrochemical, plants etc, enough pollutants can be generated to contaminate the air, land and sea around such plants.

Pollution Control Methods

The most effective method so far has been legislation backed by scientific data. Once a substance has been found to be toxic or environmentally dangerous, an effective ban or imposition of an emission limit usually sets the polluting company going towards a solution. The next most effective method is technical processing. It is usually expensive. The method of processing depends on whether the substance is gaseous, liquid or solid.

Gaseous pollutants are dealt with by scrubbing, electron beam, or filtering. Liquids can be treated by oxidation, reduction, neutralisation etc. Inert solids can, additionally, be treated by encapsulation and burial in underground sites. The newest and, eventually, the most effective method is the method of building in waste minimisation right from conception through design to commissioning of the plant. It involves consideration at each stage of process development and design the alternative issues of process changes, recycling, source treatment and administrative controls such that the manufactured product comes out at the lowest cost with the least possible pollution.

Conclusion

Pollution is essentially waste. The chemical industry is potentially the biggest environmental polluter. It has also been the most responsive and responsible in recent times in controlling pollution. A key element in pollution control is legislation and this is recommended in tackling the growing pollution problem in Nigeria.

2.5: Chemical Engineering Prospects In A Depressed Economy

A paper presented at the 1992 Annual Seminar of the IMT Student's Chapter of the Nigerian Society of Chemical Engineers, on July 15th, 1992.

Abstract

The prospects of chemical engineering in a depressed economy have been discussed. The depressed economy, in Nigeria now, is characterised by a decline in aggregate demand, low capacity utilisation and high unemployment and it persists in spite of efforts by several Nigerian governments to improve it, because of the structure and operation of the Nigerian economy. Suggestions for coping with the situations have been presented.

Introduction

The economy of nations, institutions and individuals has always attracted more than passing interest especially during hard-times. It did not come as a surprise to me, therefore, that students, preparing for a career in chemical engineering, in 1992, in Nigeria, would want a talk on the prospects of chemical engineering in a depressed economy.

The Nigerian economy is indeed depressed. The 1992 budget speech described the economy as follows:

> "The performance of the economy in 1991, as measured by the most critical macro economic indicators, showed that there was a slow down in the rate of growth of the economy compared with that of the preceding year. The real gross domestic product (GDP) grew by 4.3% in contrast to 8.2% achieved in 1990. There were indications of a reversal of the improving trend in the rate of inflation and the depreciation of the exchange rate of the Naira, that started in the

second half of 1990. Also there was a marked increase in the composite unemployment rate by the end of June, 1991" (Nigerian Trade Journal '92).

The words are different this year from those of last year but the message has been the same, year after year since the 1980s. It is thus easy to be pessimistic and to say that chemical engineering prospects must be bleak under these conditions. The temptation, of course, is not to dampen your hopes and to say that, come what may, chemical engineering is a great discipline that brooks no despair and will triumph no matter what

I will not succumb to any of these temptations in this paper. Rather I intend to follow the narrow path of objective analysis of available facts in the search for balance, and above all, for enduring truth. For it is the truth, years after you have made and implemented your choices, which remains constant.

Depressed Economies

The subject of economics and economists have always been the butt of many jokes. This is, indeed, a very welcome relief to the very serious consequences attendant upon wrong economic decisions. According to the more popular jokes, economics is the dismal science. According to another, economists are always two-handed especially in times when you need a one-handed economist. Worst of all, economists never agree with each other. But it is economics and economists that, among the many things they do, define the indicators by which the health of an economy may be assessed.

According to them, a depressed economy is characterised by a decline in aggregate demand of goods and services, low capacity utilisation and high unemployment. It is easy to see why these parameters are indicators of the health of the economy. The aggregate demand of goods and services consists of consumption, investments and government expenditure. It is not difficult to see that, in a healthy economy, there is greater consumption of goods and services and consequently greater incentives for investments and expenditures in order to produce those goods and services. This, of course, means greater utilisation of production capacity and more employment.

In Nigeria, the problem is that the measure of aggregate demand, the gross domestic product, is not only consistently low per capita (N1071.36 or about 60 US Dollars), but also capacity utilisation continued to fall from a low of 37% in 1990 to 29.8% in 1991 while unemployment continued to increase from 3.5% in December 1990 to 4.2% in June 1991 (a 20% increase in 6 Months.) (Ibid, 1992).

How Long the Depression?

One of the long lasting disputes in economic theory has been that between the supporters of the free market economy and those in favour of planned economies. Although their theoretical bases lie in one or more of the economic postulations of Adam Smith (*The Wealth of Nations*), Maynard Keynes (*The General Theory of Employment, Interest and Money*) and Karl Marx and Frederich Engels (*Das Capital*), in practice three mainstream models have emerged namely the strictly socialist economies of the former USSR, which have now collapsed, the neo-socialist economies of third world countries, which have also failed, and the free enterprise capitalist economies of Western Europe and the Americas which though apparently never at rest, is resilient and so far, the most successful.

In thus discussing economic depression in Nigeria, it is not just enough to have the figures. We must know the reason behind the figures. Take the GDP for example. When looked upon in terms of expenditure, it consists of three main items (Bowen, 1979)

- government spending at the federal, state and local government levels.
- private investment in business and housing
- personal consumption in durables and non-durables including food and various services.

In terms of income, it consists of:-

- capital allowances, corporate profits etc
- indirect taxes, interest, proprietors and rental income
- wages and salaries.

In free enterprise economies, private investment and personal expenditure on the expenditure side, capital consumption and wages and salaries on the income side constitute the bulwark of the economy while government expenditure or taxes provide guiding or stabilising influences.

In Nigeria and other third world countries, it is the other way round. Government expenditure supersedes everything that even the nature of capital consumption is unduly influenced by government. Because government is more inefficient than the private sector in resource utilisation, it is no surprise then that a depressed or inefficient economy will result in such a situation as Nigeria.

The figures on capacity utilisation are indeed misleading. Recorded industrial capacity in Nigeria is irrelevant, in the main, to a self sustaining Nigerian economy. All the production capacity of the industries in the first and second tier of the Nigerian Stock Exchange is geared to consumption of imported raw materials which is to be financed with proceeds from the sale of oil. Thus heads or tails we lose, high or low capacity utilisation we lose.

What is Employment in Nigeria?

More than 80% of employable persons in Nigeria are self employed mostly in the rural areas in subsistence agriculture and part-time artisanship and, in the urban centers, in disguised unemployment, as artisans, traders, transporters etc. Of the about 20% in regular employment, more than three quarters are employed in the public service, largely in unproductive or under-productive capacities, while the other quarter, employed in the formal private sector, service foreign interests propagated by this sector. Thus our depressed economy has a long life expectancy especially if nothing is done before our oil supply runs out, if current predictions are right, in twenty to thirty years time.

What to Do?

It is not as if various Nigerian governments are unaware of this problem or have done nothing about it. Post-Independence, pre-Nigeria civil war governments sought to solve the problem in a gradual manner by the use of a mix of import substitution industries, manpower substitution and elite substitution strategies so that over a period, the requisite industrial and productive capacity, work force and management cadre for a self sustaining economy would be installed. But they underrated the effect of intellectual under-development in the country so that political solutions were applied to all problems, political or not, with disastrous consequences. As in all under-developed societies, wealth redistribution became more important than wealth generation in spite of the fact that there was really no wealth to distribute and all the parameters necessary for equitable wealth re-distribution were either unknown or barely understood.

Post Nigerian civil war government relied entirely on political solutions and the throwing of money at every thing. When the spending stopped without much to show for it, the Obasanjo and Buhari federal governments sought for living within our means and tightening of belts in a culture where the wealth redistribution syndrome was still dominant. Shagari thought that wider and more ethical participation would do the trick but political solutions still dominated the landscape. Thus the Babangida administration sought control of the political landscape in order to implement technical and other solutions. Yet these implementations must at best be half hearted and cosmetic if political control is to be maintained at least till December 1992.

The Role of Chemical Engineering

So, where does chemical engineering come into all these? How do all these affect you as a person preparing to enter the market place of work, industry and commerce?

I shall deal first of all with chemical engineering. In a previous publication (Nnolim and Onwuka, 1981), it was claimed that chemical engineering was indeed the mainstay of western civilisation since its chemical and allied process and manufacturing industries provided the essential necessities of life and enhanced the well being of their populace. Being concerned with large scale processing of materials, it cooperated with all professions and disciplines to satisfy basic and vital human requirements in food, clothing and shelter, health and hygiene, transport and communication, culture, education, leisure, power and energy.

Chemical engineering is able to do all this because it seeks, as a discipline, a thorough understanding and familiarity with all disciplines connected with its central objective of developing, operating and managing systems, whether discrete or integrated, for large scale processing of materials. To be able to do this, however, chemical engineering requires well trained, industrious, and dedicated practitioners. It requires an investment climate that is neither too liberal nor too restrictive. And it requires transparency in all its interactions. These unfortunately, do not exist in significant proportion in Nigeria today. And this is where I come to you who are students just about to enter today's market place. What are your prospects?

In paid employment, your chances are a bit better than those of other disciplines since there is a greater demand for chemical engineering expertise in Nigerian public and private industry. Absolute demand, however, is low because of low capacity utilisation so that no company will employ more persons to man idle capacity. This, however, does not mean that you still cannot get a job.

In self employment your chances are fifty-fifty. Even when the going was good, investment capital, though abundant, was extremely difficult to get for new entrants into the industry. In these hard times, investment capital is almost out of the question for almost anybody. It is advisable for you not to bank on obtaining immediate assistance under the National Directorate of Employment (NDE) and National Economic Recovery Fund (NERFUND) since their processes are slow and tortuous and their reach extremely limited.

This, also, is not to say that you cannot obtain investment capital to start a new business. A route that is open to you and has helped a number of our past student is to seek employment outside traditional chemical industry such as teaching, banks, immigration and customs departments, forestry, agriculture, marketing and commerce and construction industry, etc. We have found, from those that we know about, that the superior training received both as chemical engineers and from our Department has put them ahead of their contemporaries in these endeavours. Another route open to you is self employment by utilising the project ideas or skills acquired in your ND or HND projects, to provide marketable equipment designs, products or services at an initial level which you or your guardian or friends can afford to finance.

Conclusion

The Nigerian economy has been depressed since the 1980s and will continue to be so until private investment and initiative are installed in the key sectors of the economy. This will happen when it is realised that political solutions are not a panacea to all national problems. Chemical engineering has a role in the improvement and sustenance of national economies but only if the requisite investment climate and manpower effectiveness, operational flexibility and transparency are in place or encouraged. New chemical engineering graduates entering the market place may seek scarce jobs in both traditional and non-traditional sectors of the economy. Or they may seek self employment through provision of equipment design, products and services in the small scale business sector.

References

1. Bowen W. (1979) The Decade Ahead: Not so Bad, if We do things Right; Fortune, Vol. 100 No. 7, Oct. 8, Time Inc., Chicago, USA.
2. Nnolim, B.N. and Onwuka N.D; Chemical Engineering Key to Industrial Development in Nigeria, Nigerian Society of Chemical Engineers, Anambra State Branch, 1981.
3. "Watershed of Our National Evolution". The 1992 Budget Speech; Nigerian Trade Journal Vol. 36 No. 1 Jan - March 1992, pp 6 - 23, Prodn. and Film Dept. Federal Ministry of Information, Ikoyi Lagos.

www.ingramcontent.com/pod-product-compliance
Lightning Source LLC
Chambersburg PA
CBHW031537210526
45464CB00003B/1054